SHARK FILES

MAKO SHARKS

by Julie K. Lundgren

A Crabtree Seedlings Book

TABLE OF CONTENTS

Crabtree Publishing
crabtreebooks.com

School-to-Home Support for Caregivers and Teachers

This book helps children grow by letting them practice reading. Here are a few guiding questions to help the reader with building his or her comprehension skills. Possible answers appear here in red.

Before Reading:

- What do I think this book is about?
    - *I think this book is about mako sharks.*
    - *I think this book is about what mako sharks like to eat.*
- What do I want to learn about this topic?
    - *I want to learn where mako sharks live.*
    - *I want to learn how many teeth mako sharks have.*

During Reading:

- I wonder why...
    - *I wonder why mako sharks have pointed noses.*
    - *I wonder why anyone would think it was fun to catch a mako shark.*
- What have I learned so far?
    - *I have learned that mako sharks have long, sleek, strong bodies.*
    - *I have learned that they swim very fast.*

After Reading:

- What details did I learn about this topic?
    - *I have learned that mako sharks hunt fast prey like tuna and swordfish.*
    - *I have learned that mako sharks can leap to reach food.*
- Read the book again and look for the vocabulary words.
    - *I see the word* ***torpedoes*** *on page 13 and the word* ***endangered*** *on page 21. The other glossary words are on pages 22 and 23.*

MAKO SHARKS

Meet the world's fastest shark.

Mako sharks are built for speed.

FROM THE FILES

They swim about 10 times faster than the fastest human swimmers.

They have sleek, strong bodies.

FROM THE FILES

Pointed noses cut the water.

They can swim across entire **oceans**.

They hunt fast **prey** such as tuna, **swordfish**, and other sharks!

tuna

swordfish

Mako sharks can leap to reach food.

FROM THE FILES

Some call them **torpedoes** with **teeth**.

Their teeth are long and sharp.

People catch them for sport and food.

How many makos are in our oceans? We do not know.

We need to learn more about these speedy hunters.

FROM THE FILES

Mako sharks are **endangered**.

GLOSSARY

endangered (en-DAYN-jurd): Endangered animals are animals that are in danger of becoming extinct.

oceans (OH-shuhnz): Oceans are large bodies of saltwater where many animals live.

prey (PRAY): Prey is any anima hunted and eaten by another animal.

swordfish (SORD-fish): Swordfish are large fish with very long, pointy noses.

teeth (TEETH): Teeth are white, bony parts inside a mouth that are used for biting and chewing.

torpedoes (tor-PEE-doze): Torpedoes are fast, sleek missiles launched underwater.

Index

About the Author

Julie K. Lundgren

Julie K. Lundgren grew up near Lake Superior where she reveled in mucking about in the woods, picking berries, and expanding her rock collection. Her interests led her to a degree in biology. She lives in Minnesota with her family.

Websites

www.dkfindout.com/us/animals-and-nature/fish/mako-shark
https://easyscienceforkids.com/all-about-sharks

Crabtree Publishing

crabtreebooks.com 800-387-7650

Written by: Julie K. Lundgren
Designed by: Jennifer Dydyk
Edited by: Kelli Hicks
Proofreader: Melissa Boyce

Photographs:
Shark illustration on cover logo © BATKA/Shutterstock; white shark illustration for "FROM THE FILES" © Dashikka/Shutterstock; Cover © Alessandro De Maddalena/Shutterstock.com; page 3 © wildestanimal / Shutterstock.com; page 5 © Martin Prochazkacz/Shutterstock.com; page 6 © Martin Prochazkacz/Shutterstock.com; page 7 © wildestanimal/Shutterstock.com; page 9 © Martin Prochazkacz/Shutterstock.com; page 11 tuna © Shane Gross/Shutterstock.com, swordfish © SVITO-Time/Shutterstock.com; pages 13, 17 and 19 © Alessandro De Maddalena/Shutterstock.com; page 15 © saulty72/Shutterstock.com; page 21 © saulty72/Shutterstock.com; page 23 torpedo © Ilya Shulika/Shutterstock.com

Hardcover	978-1-4271-5848-2
Paperback	978-1-4271-5849-9
Ebook (pdf)	978-1-4271-5850-5
Epub	978-1-4271-5851-2
Read-along	978-1-4271-5852-9
Audio book	978-1-4271-5853-6

Printed in Canada/072023/CPC20230728

Published in Canada
Crabtree Publishing
616 Welland Avenue
St. Catharines, Ontario
L2M 5V6

Published in the United States
Crabtree Publishing
347 Fifth Avenue
Suite 1402-145
New York, NY 10016

Library and Archives Canada Cataloguing in Publication

Available at the Library and Archives Canada

Library of Congress Cataloging-in-Publication Data

Names: Lundgren, Julie K., author.
Title: Mako sharks / Julie K. Lundgren.
Description: New York : Crabtree Publishing, [2022] | Series: Shark files - a Crabtree seedlings book | Includes index.
Identifiers: LCCN 2021018456 (print) | LCCN 2021018457 (ebook) | ISBN 9781427158482 (hardcover) | ISBN 9781427158499 (paperback) | ISBN 9781427158505 (ebook) | ISBN 9781427158512 (epub) | ISBN 9781427158529
Subjects: LCSH: Mako sharks--Juvenile literature.
Classification: LCC QL638.95.L3 L865 2022 (print) | LCC QL638.95.L3 (ebook) | DDC 597.3/3--dc23
LC record available at https://lccn.loc.gov/2021018456
LC ebook record available at https://lccn.loc.gov/2021018457